Workbook, Online
Courses Available

HOW TO CRUSH SOCIAL MEDIA IN ONLY 2 MINUTES A DAY

TWITTER, FACEBOOK, INSTAGRAM,
KRED, GOODREADS, LINKEDIN

Ndeye Labadens

How to Crush Social Media in Only 2 Minutes a Day

Youtube, Google, Amazon, Cross Promotion, blogs, and Shapr

Workbook 1

Ndeye Labadens

Signup for a free ebook here

lannconsultings.com

OTHER BOOKS BY NDEYE

Australian Memories - Discover Aussie Land and the Mysterious Red Center

Relocation Without Dislocation: Make New Friends and Keep the Old

Secrets Book Launch Journey to the Ultimate Success Book 1-3

African Memories: Travels to the interior of Africa

European Memories: Travels & Adventures Trough 15 countries

Playbook Strategy: Stand Out Like a Business Giant

Asian Memories: The Delights and Discoveries of an incurable traveler

Secrets Book Launch Journey to the Ultimate Success Book 4 - 6

How to Crush Social Media in Only 2 Minutes a Day: Twitter, Facebook, Kred, Goodreads, LinkedIn

Coming soon!

American Memories: Cross Continent Adventures and Discoveries from North to South

ACKNOWLEDGEMENTS

There were so many awesome people who inspired me to write this book, and helped me continually improve it.

Thanks for your support, and for helping making this book great.

I keep learning every day, and I will be sharing my knowledge acquired with you.

Follow me on Amazon and Twitter to be informed about my next books.

CONTENTS

Preface

This workbook will help you process, understand, use and master the material you studied in 'How to Crush Social Media Marketing in Only 2 Minutes a Day: Youtube, Google, Amazon, Cross Promotion, Blogs, Shapr Book 2,' found at this Amazon.com link:

https://www.amazon.com/Crush-Social-Media-Only-Minutes-ebook/dp/B07FDWXMRP

It is to your benefit to study the key points made in the book which will go a long way to helping you crush social media marketing. A good learning technique is to tell the student what will be taught, and then teaching it, and then telling the student what was taught. That is the purpose of this workbook.

'How to Crush Social Media Marketing in Only 2 Minutes a Day: Youtube, Google, Amazon, Cross Promotion, Blogs, Shapr Book 2,' taught you the basics, and now you will review what you were taught. Your investment in yourself by evidence of purchasing this workbook will pay off royally when you put your knowledge to work. When you work it, it will work for you giving considerable return. Online marketers are doing this same thing every day and receiving financial rewards. Consistency, with adaptability when called for, is the key attribute for your success. Some people fear their own success. Is that you?

At the risk of sounding tedious, we repeat and reemphasize some of the material you learned, but it is only due to our earnest desire that you understand and get the full benefit of, 'How to Crush Social Media Marketing in Only 2 Minutes a Day: Youtube, Google, Amazon, Cross Promotion, Blogs, Shapr Book 2.' We do it to give you practical advantage over your numerous online marketing competitors.

As you already know from the book, connecting with your target market will help you get more subscribers. More subscribers and customers equal more potential sales for you. Without customers and sales, you will need to question whether or not you have a business.

P. Sterling Underwood

https://www.amazon.com/P.-Sterling-Underwood/e/B005FCC9BG

Introduction

Social media marketing is a process for gaining attention and world wide web traffic through the social media web sites.

1. Social media is a very strategic marketing platform (check your choice).

☐ True   ☐ False   ☐ I don't know

2. The social media platform is the perfect vehicle to reach the right audience and its masses of people—different cultures, ages, religion, sexes, locations, interests, etc.(check your choice).

☐ True   ☐ False   ☐ I don't know

3. To crush social media marketing, targeting your market is key and best use of your time and energy (check your choice).

☐ True   ☐ False   ☐ I don't know

4. If unsure of any answers thus far, read my Amazon Kindle ebook, 'How to Crush Social Media Marketing in Only 2 Minutes a Day: Youtube, Google, Amazon, Cross Promotion, Blogs, Shapr Book 2,' found at this link:

https://www.amazon.com/Crush-Social-Media-Only-Minutes-ebook/dp/B07FDWXMRP

because each statement is positively true.

5. Publicity for your product or service through a trusted third-party resource is proven to create interest in and credibility for your product and/or service.

☐ True   ☐ False   ☐ Not Sure (Not sure? Go to question #4)

6. Promotion by trusted third-party resources gives your business cover and extends your reach beyond your anticipated audience?

☐ True   ☐ False   ☐ Not Sure (Not sure? Go to question #4)

7. Circle which statement is true. "Every online marketer needs_____":

A) a goal,

B) a product,

C) a service

D) a cause

E) all of the above

Correct answer: E--All of the above.

8. Using the social media platform is the perfect vehicle for targeting and reaching the right audience to achieve total success.

☐ True   ☐ False

Correct answer: True.

9. Leveraging social media for achieving success means focusing on applicable demographics—group ages and other factors—relative to your product or service which might result in your videos and news going "viral."

☐ True   ☐ False

Correct answer: True.

10. The social media world is wide and more extensive than ever.

☐ True   ☐ False

Correct answer: True.

Social Media Sites

Most social media sites support seamless integration to various devices such as a portable PC, desktop, tablet, and mobile phone.

Facebook

1. Facebook has lots of features which help you make money.

   ☐ True   ☐ False

   Correct answer: True.

2. Facebook's features can publicize and promote your wares.

   ☐ True   ☐ False

   Correct answer: True.

3. A Facebook business page is recommended as a free tool which helps you interact with prospects and customers.

   ☐ Agree   ☐ Disagree

   Correct answer: You should agree.

4. A Facebook business page can improve your business with thousands of users in your area.

   ☐ True   ☐ False

Correct answer: True.

5. Check which feature(s) is provided by a dedicated Facebook business page:

☐   direct interaction with prospects and customers via your posts

☐   uploading pictures, products, and videos of your product or service

☐   help build a database of people that share your posts

☐   commuting Facebook users stay connected to you allowing round the clock go anywhere   promotion of your attractive and worth-checking-out products and services of interest

☐ news feeds reach prospects and customers expediently

If you checked all five features, go to the head of the class. You have learned aell. Congratulations!

6. Which marketing tool below will help your business connect with your customers—big and small?

☐ Having a Facebook business page

☐ Using a Facebook business page to advertise to prospects and customers

☐ Using a Facebook business page to target specific customers through behaviors, demographics, and interests

If you checked all three options, great! You show great business sense.

7. Can a Facebook business page boost your business if your business lacks one?

☐ True   ☐ False

Correct answer: True.

Blog/Blogging

Quality of content is a hallmark of good blogging. Think of a blog as awaited news. It is a semi-professional way to address the interests of your audience by timely, useful, and well-written communications. A good blog delivers meaningful news to prospects and customers—news they need, news they want to read, and news they will perhaps soon use.

1. Consider reasons to blog that will help ignite your social media marketing presence. Think about how trending topics you stay aware of can tie into social media marketing for your product and service. Jot down a few ideas (from current headlines, topics, links, etc.) that have tie-ins to the product or service you offer to prospects and customers.

1. _____   3. _____

2. _____   4. _____

2. Which statement is NOT true for blogging:

   A) Content Management Services (CMS) are free services to get your blog up and running

   B) Examples of a CMS are Blogger, WordPress and Tumblr

   C) A kinship with your audience is key to writing a well-received blog

   D) Keyword-driven ranking is unnecessary in order for people to see your blog

Correct answer: 'D--Keyword-driven ranking is unnecessary for people to see your blog is NOT A TRUE STATEMENT. 'D' is the answer to the question.

3. Complete this sentence. Search Engine Optimization (SEO) is worth knowing about because:

_____

___.

Answer: If you wrote something like: "SEO is Search Engine Optimization knowledge, which is essential and instrumental for achieving goal success as an online marketer,"

you understand that SEO knowledge is worth knowing about and understanding.

4. Is this a true statement? The right keywords will improve your SEO ranking for your blog.

   □ True   □ False

Correct answer: True. (The keywords you decide on should relate to your blog posts as much as possible in order to be ranked as high as possible by the search engine, i.e., Google, Bing, etc.)

According to my Amazon Kindle ebook, 'How to Crush Social Media Marketing in Only 2 Minutes a Day: Youtube, Google, Amazon, Cross Promotion, Blogs, Shapr Book 2,' found at this link:

it is wise to use a keyword search tool to guage competition and the number of searches for a given keyword. The fewer the competitors and higher the number of searches for the keyword in a month, the more advantageous it is for you to go with that keyword. You want to make your selected keywords very specific so people searching for your product or service find you first.

5. If you want your blog or website to be found with ease by those looking for your specific product or service, which following technique(s) should you choose:

A) Conduct SEO research using a keyword search tool

B) Use keywords that pop into your head without vetting whether they are your smartest choice

C) Add additional keywords to the results of your keyword search tool investigation which go straight to the point of the specific services your product offers

D) Hope your blog shows up on page one of SEO search results

Answer: A, B, and C. (Never D.)

Social Media Integration In the Blog Space

Social media integration is what you do to make sure your blog/website existence gets in front of the inquisitive eyes you want knowing about your product and/or service. It is a thing you do purposely.

CMS Buttons

There are many options which facilitate sharing the content of your blog/website. Social media buttons provided by CMS' (Content Management Services) like Tumblr are your friends.

This is what you do and how you do it.

Within the CMS, look for social media 'buttons' which turn on 'Share,' 'Like' and 'Dislike:'

1. Enable the options if not already enabled by default.

2. Arrange settings so that every post will show buttons for 'Share on Tweeter (??), or Google+, or Facebook, etc, and Re-blog within the blogging network you are affiliated with

3. Create quality, eye-catching content

4. Leverage outstanding opportunities to reach a wider audience

Twitter

Twitter use is wide-ranging, popular with businesses, celebrities, musicians, actors, everybody!

Name one thing which accounts for this platform's extreme success.

_____

If you thought mass appeal, or writing and posting to specific followers or groups, or instant responses, or bantering interactively with specific followers or groups without incurring email delays, or hands-on management of advertising and marketing campaigns regarding brands and products, go to the head of the class because you got it right.

From what you know about your own Twitter use, what other benefits can you identify?

_____   _____   _____   _____   _____

How about this one: The 140 character limitation has been increased, allowing more space to get your point across.

With 140,000,000+ Twitter users worldwide, what would say is its chief advantage?

_____

If you replied anything similar to the following, you elevate your potential for online marketer success:

"It implies immense potential to boost awareness of my product and/or services. I can expect to grow my customer base substantially with economy of effort and expense."

Ask yourself whether it is worthwhile using Twitter? Does this make sense, agree or disagree?

"With Twitter, I can reasonably expect my tweets to engage people with my brand, make them aware of it, get them talking about it, reviewing it and telling others about my events, broadcasts and various promotions."

Correct answer: You should agree. It is a quite reasonable expectation.

LinkedIn (i.e., Linked In)

LinkedIn is a popular social network platform primarily used for professional networking, including employers posting jobs and job seekers posting their CV/Resumes. It is a business and employment-oriented service that operates via websites and mobile apps. Business professionals and entrepreneurs use it to mingle with like-minded associates for strategic career and basic networking engagement. There is nothing strictly social about it as is the case with Facebook, Twitter, Instagram, etc. Outside the business realm, it is less well known and little used. LinkedIn is intended to filter and leave the fun behind to focus deeper on professional contacts, mining business and job opportunities across the social media world.

1. If you intended to float career-related information about yourself on the World Wide Web, specifying personal information about your education, studies, interests, contact information, certifications, degrees, share information about your brand, service, product, etc. the the most-welcoming place to do it in the social media world would be:

A) Facebook

B) Tumbler

C) WordPress

D) LinkedIn

Correct answer: D--LinkedIn, because that's where serious-minded professionals hang out.

2. You can create a personal profile and/or a business/company page on LinkedIn just like you can on Facebook. True or False?

Correct answer: True.

3. Linkedin is a great place to keep your audience and followers up to date with the latest information about your product and/or service. True?

Correct answer: True.

YouTube

People find your vids (videos) on YouTube one way or another, and that is why you want to consider the huge number of eyeballs that prefer viewing social media content over reading social media content. It is accurate to say that interests lead people around by the nose. Where interests lead, people follow, and where people end up is where you want to be for them to find links to your blog and websites. YouTube allows users to upload, view, rate, share, add to favorites, report, comment on videos, and subscribe to other users. Most of the content on YouTube is uploaded by individuals, but media corporations including CBS, the BBC, Vevo, and Hulu offer some of their material.

1. After you have created and uploaded your marketing video to YouTube, through which gateway is a web visitor likely to find your video?

A) By being redirected by someone else's website to their own video on YouTube where the title of your video conveniently shows up as a 'Related video' to click on

B) You show up in search engine results along with websites and blogs

C) By convenient backlinks redirecting to your YouTube video from your own website(s)

D) YouTube visitors clicking through videos to discover your video featured in related categories

E) You set up a YouTube channel in support of achieving your goal, product, service and cause as an online marketer, and people subscribe to it

Correct answer: A, B, C, D, and E, all of the above.

2. YouTube is the social media site where you show, not just tell. If you had the need, would YouTube be the appropriate platform to demonstrate how to use your product? Why is this a great idea?

_____

If you wrote something similar to:

"It offers a wide variety of user-generated and corporate media videos. Available content includes video clips, TV show clips, music videos, short and documentary films, audio recordings, movie trailers, live streams, and other content such as video blogging, short original videos, and educational videos, so, why not use it to demonstrate use of my product?"

Then, you got it right.

3. Does it make solid business sense to redirect your YouTube video viewers to your business website where a purchase can happen?

Correct answer: Yes, indeed.

4. As a smart business person and online marketer exploiting your online marketing social media s,[ltrategy to reach every single corner of the internet, what do you think is the best use of YouTube for accomplishing what y vbou have set out to do?

A) Product/Service Announcements

B) Product/Service Demonstrations

C) Socializing

D) Producing 'How-to' videos

Correct answer: A, B, and D for sure. (Socializing....??? Only, if socializing for a cause).

Google+

As of this writing, the best to be said about Google+ is that it is a good idea for an online marketer to have a spare Google+ account in his or her back pocket. No source is too little or too much in social media marketing because extended reach is the engine for online marketing. That being said, its popularity should pick up because its exclusivity and integration of services has strong appeal for professionals and business networks.

1. Which service is not provided by Google+:

A) Gmail (Google's email service)

B) 'Stream' feature similar to Facebook's News Feed

C) Status sharing (Google+Basics, Google+Circles)

D) Access to the whole Google+ network of services via Gmail account creation

E) YouTube

F) PayPal

G) Google Search Engine

Correct answer: F. (A, B, C, D, E, and G are Google+ services)

Social Media Marketing for Business

As stated in 'How to Crush Social Media Marketing in Only 2 Minutes a Day: Youtube, Google, Amazon, Cross Promotion, Blogs, Shapr Book 2,' found at this link:

https://www.amazon.com/Crush-Social-Media-Only-Minutes-ebook/dp/B07FDWXMRP,

social media marketing is a powerful way for businesses of all sizes to reach prospects and customers.

Rather than hope for the best results by your bland, uninteresting, stagnate online presence on the World Wide Web, social media marketing is your best bet for:

- increasing website traffic and

- building conversions and

- raising brand awareness and

- creating a brand identity for positive brand association and

- facilitating communication and interaction with key audiences.

1. Social media marketing is a powerful way of speaking directly with your business audience.

☐ True   ☐ False

Correct Answer: True.

2. Examples of social media platforms for reaching out to your business audience are Facebook, Twitter, Instagram and Pinterest.

☐ True   ☐ False

Correct Answer: True.

3. Marketing via social media platforms can bring remarkable success to your business because of its potential to:

A) create devoted brand advocates

B) drive business leads

C) drive sales

D) All of the above

Correct Answer: D--All of the above

4. Social media marketing, or SMM, is a powerful online marketing tool.

☐ True   ☐ False

Correct Answer: True

5. Social media marketing utilizes proven internet marketing methods to achieve your marketing and branding goals by:

A) content creation and sharing on social media networks

B) posting text, and image updates and videos on social media networks

C) driving audience engagement

D) paid social media advertising

E) All of the above

F) Only A and C

G) Only B and D

Correct Answer: E—All of the above.

6. State the first action to take before/prior to starting your social media campaign?

_____.

Correct Answer: Start with a plan.

7. You start with a plan because: _____.

Correct Answer: Because you need to define (or, map out) how you are going to reach your business goal. Your Social Media Marketing Plan should address:

- What (the end goal accomplishment of your planning effort)

- Who (identifies the target audience to whom you plan to market and sell)

- Where (on the World Wide Web you expect to locate your adoring audience)

- Your Message (to immediately grab the attention of your prospects)

- Your Business Type (Your business type should inform and drive your social media marketing strategy.)

8. The bigger and more engaged your audience through social media networks, the easier to   achieve your marketing goals.

☐ True   ☐ False

Correct Answer: True

9. A Facebook page helps you connect with your customers.

☐ True   ☐ False

Correct Answer: True

9.Setting up your own Facebook business page yourself is easily done and recommended.

☐ True   ☐ False

Correct Answer: True

Great content rules when it comes to social media marketing. Knowing your competition   matters just as much. Therefore, building a well-designed social media marketing plan is   essential for conducting successful online marketing.

10. Name the two types of research recommended in, 'How to Crush Social Media Marketing   in Only 2 Minutes a Day: Youtube, Google, Amazon, Cross Promotion, Blogs, Shapr Book   2,' found at this link:

https://www.amazon.com/Crush-Social-Media-Only-Minutes-ebook/dp/B07FDWXMRP

that will help you better structure your social media marketing strategy:

A) Keyword research

B) Competitive research

C) Statistical research

D) Trend Analysis

Correct Answer(s): A, B, Keyword and Competitive Research

11. The purpose for competitive research is:

A) copying your competitors' content as a shortcut to creating your own

B) discover what other businesses in your industry are doing to drive social media engagement

C) to brainstorm ideas that will interest your target audience

D) None of the above

Correct Answer(s): B and C. B--Discover what other businesses in your industry are doing to   drive social media engagement, and C--Brainstorm ideas that will interest your target audience.

12. What are key practices for good social media marketing:

A) post regularly

B) provide helpful and interesting content

C) offer truly valuable information for your ideal customers' consumption

D) include pertinent social media images, videos, infographics, how-to guides and more

Correct Answer(s): A,B,C, and D

A brand image presented with consistency across all your chosen social media platforms helps   project a recognizable core identity to which prospects and customers become familiar and   relate. While each platform has its own unique environment and voice, your business's core   identity, whether it's friendly, fun, or trustworthy, should stay consistent.

13. Content Marketing and Social Media Marketing benefit each other because they enable:

A) sharing your best site and blog content with your readers

B) building a loyal following across social media platforms

C) isolating your great content by a wall between domains

D) directing your readers to your new postings on other platforms

E) promoting your great blog content to build more followers

Correct Answer(s): A,B,D, and E. C—Isolation is not good for your self-promotion.

Sharing curated links is a good practice to adopt. It is helpful to your cause. The operative   word is 'curated' because:

- you vetted content offered by outside/external sources and want your audience to know about great, valuable information available from others besides yourself

- it is opportunity to leverage your own unique, original content to gain followers, fans, and devotees by becoming a trusted source

- curating and linking to outside/external sources elevates your trust and reliability with your followers

- you might even get some links (and followers) in return from the sources you link to

14. Which of the following provide reason to track what is going on with your competitors:

A) keeping an eye on competitors can provide valuable data for keyword research

B) keeping an eye on competitors can provide useful social media marketing insight

C) if competitors gain competitive advantage from certain social media marketing channels or techniques, you might want to consider doing the same thing

D) you want ammunition for name calling

Correct Answer(s): A,B, and C.

How to Make Money on YouTube

Conducting electronic marketing employing YouTube videos requires strategic thinking because it is a way to generate money. The key to money-generating income is through YouTube video advertising, personal appeal videos, business presentation, or historic event clips. It is all about advertising effectively, creating the type of video that pops out from the clip once it is played. It is easy.

Partnering

There are hundreds of thousands of potential partners earning good money on their own all over the world which are accessible to you on YouTube.

1. Why is it a great idea to find someone on YouTube to partner with?

   A) As a partner, you will be allowed to put your ads on your partner's video website(s).

   B) You may get a share of revenue from partner's viewers when they click on your ads.

   C) It is a trivial matter to become a partner with someone on YouTube.

   D) Your partner does all the work for you.

   Correct Answer(s): A and B only. C is not correct because it is NOT trivial to become a partner. D is not correct because you need to learn how partnering on YouTube works.

2. True or false: To get ready for collecting YouTube revenue as a partner, you must set up an an Adsense account.

   □ True   □ False

Correct Answer: True

3. Clicks from ads on lots of different videos will pump money into your Adsense account.

□ True   □ False

Correct Answer: True

4. After establishing your Adsense account, you need to know the profile of the audience to   which you want to advertise. The best way to determine this is by matching your ad to the   video designed to reach your target audience

□ True   □ False

Correct Answer: True

5. Trending political or social topics draw attention. The video you attach your ad to should:

A) relate to political or social topics in an original, interesting way

B) not relate to current events in any appreciable way

Correct Answer: A

6. The YouTube Partner Program specifies requirements to be approved as a YouTube partner.   Which is the correct answer?

A) This is a popular program and approvals happen quickly

B) This is a popular program and approvals take time to clear you through the process

C) Impatience with the approval process is recommended

Correct Answer: A

The Power of Video Marketing

1. What type of marketing is twice as powerful AND twice as effective as banner or pop up ads?

    A) Banner ads

    B) Pop up ads

    C) a YouTube marketing video

Correct Answer: C

2. The optical length of time for a YouTube ad video to run is:

    A) 2 minutes

    B) 5 minutes

    C) 10 minutes

Correct Answer: A, 2 minutes

3. The traditional marketing model without use of videos is outdated.

    ☐ True   ☐ False

Correct Answer: True

4. Video marketing hooks the person in better than standard text sales.

☐ True   ☐ False

Correct Answer: True

5. Which statement or statements are true? Video marketing:

A) does a better job of engaging with the viewer

B) helps develop your brand

C) does a better job of getting your message out there

Correct Answer: A, B and C

6. Better ways to learn and make money with video marketing are:

A) DIY – Do it Yourself

B) Outsource to a professional who knows the ropes of video marketing

C) Sign up for a 'Make Money With Video Marketing' course

D) Make dazzling videos having special effects

E) Quickly converting an article to video and narrating over it as text rolls across
the   screen

Correct Answer: A, B and E

7. A visual presentation with audio is a better sales method than long sales copy

☐ True ☐ False

Correct Answer: True

8. Connecting with viewers with a video will help to get you more subscribers.

☐ True ☐ False

Correct Answer: True

9. Creating and uploading a video to YouTube costs nothing (except for the production cost for your own home-made video). Why should professional video companies be avoided?

A) Dazzling, Hollywood videos can distract from your message

B) Cost and production quality might might not be cost-effective

C) A simple presentation best conveys the features of your product

D) A home-made video looks more down to earth

E) A home-made video resonates better with the average viewer

Correct Answer: B, C, D and E

10. The most important aspect to a video over and above perfect picture and sound quality is:

   A) that the video engages the viewer

   B) the viewer thinks the video is so cool that the message gets overlooked

   C) the video drones on and on going for the hard sell

   D) delivering a good presentation

Correct Answer: A, D

11. What is imperative to include in your YouTube video that will help you make money and get subscribers?

   _____

Answer: Your website link

12. Where in your video should your website link be shown?

_____

Answer: It is preferable to show your website link at the end

13. Where else should your website link be found?

_____

Correct Answer: Show your website link in the YouTube description box

Email Video Marketing

1. What online marketing tool collects and retains important customer data for the online business?

    A) A word processor

    B) Statistical analysis

    C) Trend analysis

    D) An Autoresponder

2. What functions does an Autoresponder perform for the online business person?

    A) Collects and retains subscriber names

    B) Collects and retains subscriber email addresses

    C) Sends followup emails for promotional offers or general information

    D) Sends direct mail post cards

Correct Answer: A, B, C

3. Using an Autoresponder, can an email be sent to a previous subscriber with a link to your YouTube video which could be a video sales presentation for generating sales?

Correct answer: Yes. An autoresponder with video emails is an innovative approach to   crushing social media marketing

YouTube Ads

YouTube's advertising platform is one marketing strategy you should seriously consider adding to your arsenal. YouTube video ads offers cost-effective advertising options allowing you deep customization targeting very specific user groups. Multiple advertising formats and ad creation options are also available to you.

1. Do you recall what some of those customization options are for the online marketer?

_____, _____, _____

Correct Answers:

- Select target audience

- Specify budget

- Specify other essential marketing parameters

2. According to Book 2, 'How to Crush Social Media Marketing in Only 2 Minutes a Day:

Youtube, Google, Amazon, Cross Promotion, Blogs, Shapr, ' the first two steps to creating   effective YouTube video ads is to link your YouTube channel to:

A) AdWords Account

B) Autoresponder

C) CMS buttons

D) Create a new Online Video campaign

Correct Answer: A and D

3. Safe approaches to getting your campaign running and staying within budget are:

4. After initially implementing safe approaches to campaign operation, a good idea is to:

A) adjust campaign parameters depending on results

B) stick with the campaign as originally setup without parameter fine tuning

C) make changes for the luck of it

Correct Answer: A

Choosing the Right Video Ad Format for Your Business

1. YouTube video ads come in four formats. Can you name them?

Correct Answers:

1. Display ads

2. Overlay ads

3. Skippable ads

4. Non-skippable in-stream ads

2. Match the ad type to its description

   A) this type appears right of the feature video

   B) this type is semi-transparent and appears on the lower portion of your videos

   C) this type permits users to skip after 5 seconds of viewing

   D) this type must be watched in entirety before your video can be viewed

Correct Answers: A corresponds to Display ads. B corresponds to Overlay ads, C corresponds   to Skippable ads, and D corresponds to Non-skippable in-stream ads

3. Which two (2) ad types can appear before, during, or after the main video?

   A) Non-skippable ad type

   B) Display ad type

   C) Overlay ad type

D) Skippable ad type

Correct Answer: A (Non-skippable ad type), D (Skippable ad type)

4. In theory, which of the four ad types has the highest cost and is the most risky?

A) Non-skippable ad type

B) Display ad type

C) Overlay ad type

D) Skippable ad type

Correct Answer: A (Non-skippable ad type)

5. Which ad type is the most effective?

A) Non-skippable ad type

B) Display ad type

C) Overlay ad type

D) Skippable ad type

Correct Answer: A (Non-skippable ad type)

6. What is the basis for determining cost-effectiveness?

Correct Answer: Cost per thousand (CPM) impressions.

7. Which ad type is the most expensive of the four types (based on CPM)

    A) Non-skippable ad type

    B) Display ad type

    C) Overlay ad type

    D) Skippable ad type

Correct Answers: A (Non-skippable ad type)

8. What is the high risk associated with a Skippable ad type?

Correct Answer: A (high video abandonment rate)

9. Name the best ad type to initiate your ad campaign with.

Correct Answer: High priority skippable video ad

10. Name two reasons why a high priority skippable video ad type is better for initiating a campaign?

    A) Safer in terms of expense

    B) It is more entertaining

    C) You cannot turn it off

D) A more cost-effective choice

Correct Answers: A (safer in terms of expense) and D (a more cost-effective choice)

11. CPV stands for:

A) Cost Per View

B) Cost Per Voter

C) Cost Per Victory

D) Cost Per Valuation

Correct Answer: A (Cost Per View)

12. Cost Per View is your charge for:

A) the number of clicks for a viewer to skip watching your video

B) the amount you pay whenever a viewer watches a portion of your YouTube video ad

Correct Answer: B (the amount you pay whenever a viewer watches a portion of your YouTube video ad)

13. According to Book 2, the usual CPV for a locally-targeted skippable ad is between:

A) $0.35 and $0.50

B) $0.05 and $0.09

C) $0.010 and $0.30

D) $0.55 and $5.00

Correct Answer: C ($0.010 and $0.30)

14. Which CPV charge is reasonable for 'good exposure,' even on a small budget?

A) $0.35 and $0.50

B) $0.05 and $0.09

C) $0.010 and $0.30

D) $0.55 and $5.00

Correct Answer: C ($0.010 and $0.30)

15. According to Book 2, two approaches for a good bidding strategy are:

A) win a bidding war regardless of your budget

B) bid high at the outset to drive off other bidders

C) make an in-stream bid that amounts to half of your cost for 'good exposure'

D) tweak your bid depending on the performance of your ads

Correct Answers: C (make an in-stream bid that amounts to half of the cost) and D (tweak your   bid depending on the performance of your ads)

Targeting Your Audience

1. After choosing your ad type, the next step is:

   A) Define your target audience

   B) Create engaging marketing videos

   C) Open a bank account

   D) Utilize video marketing data from past campaigns

Correct Answer: A (define your target audience) and D (utilize video marketing data from past   campaigns)

2. Examples of demographics to mine from any previous campaign are:

   A) gender

   B) age

   C) combining demographics targeting with keyword targeting to focus on a specific   category of users such as men between 25-45 years who like cars and watch videos on   car cleaning and maintenance, or women between 23-56 who love birds and watch   videos about canaries

Correct Answer: C (all of the above)

Remember, placement matters. Placements enable you to target individual YouTube videos or   specific YouTube Channels, which can come in handy if you want your ads displayed in a   partner's videos. After you've chosen your audience, you can create your ad and add a banner   to it to make it stand out.

YouTube Video Ad Tips

- Keep your YouTube in-stream video ads separate from your YouTube Search ads (which display the video ad within the video search results), by targeting them to separate audiences.

- Set a low maximum daily budget to minimize unprofitable views at the beginning of your campaign and keep unexpected volume in check.

- If you're on a small budget, focus on targeting small groups while ensuring that your predicted views don't exceed 1,000.

- Smaller campaigns are easier to monitor and to adjust.

- Lower a predicted audience estimate by adding overlapping keywords.

- Retarget your ads to your YouTube viewers using the AdWords remarketing tool to increase conversion rates and improve brand awareness and recall.

How to Make Money on Amazon

Amazon is one of those sites that have constant web traffic. Because of this, you can earn an lot by knowing how to make money with Amazon. It really depends on what you are into and what your personality is.

The Amazon Advantage Program

1. The Amazon Advantage program is perfect for book promotion and readers. Which of its features allows publishers to avoid the problem of having published too many books with very little sales?

   A) On Demand Publishing Program

   B) Hit-and-miss publishing program

   C) Wiley Fox on-demand printing company

Correct Answer: A (On Demand Publishing Program)

2. Is the Amazon Advantage On Demand Publishing Program a great help for self-supporting writers?

Correct Answer: Yes. The book can be listed in Amazon after you sign up.

Amazon Associate Program aka Affiliate Program

If you are more of a salesman rather than a writer, then you may still want to know how to make money with Amazon. If you are into affiliate marketing, or eager to start trying your luck with it, sign up with Amazon's Associates program. All you have to do is prepare a web site about anything, and make sure that your page is entertaining and informative. For example, if your web page is about tennis, then you can create a store filled with books and other reading materials about tennis.

1. Since Amazon is a commercial website where one can buy, sell and see over 20,000 products both new and old stuff, how can the individual internet entrepreneur exploit it to derive income?

A) You go to their shopping mall as a consumer of products

B) You make money by promoting the products of a company in return for a commission on all sales generated by you

C) You use it to search out discount coupons

Correct Answer: B. You make money by promoting the products of a company in return for a commission on all sales generated by you

2. As an affiliate marketer, you position yourself as a middleman between the Amazon.com visitor with your own network of websites and marketing methods, and the vendor of various products. What does this gain you?

A) An Amazon affiliate account

B) A unique link to use to promote the vendor's products

C) The right to earn commission on all sales generated by you

D) All of the above

Correct Answer: D (All of the above)

3, How much commission is paid on products you sell?

Answer: As of this writing, the Amazon affiliate program pays you 50% to 75% of the revenue they generate from each product that you refer.

4. What is a high-gravity product?

Answer: A high-gravity product means those products people buy often, either daily or weekly.

5. What is a low-gravity product?

Answer: A low-gravity product means those products people do not buy often.

6. The Amazon Associate Program offers various tools that help you crush social media marketing through Amazon. Recall what you read about in 'How to Crush Social Media   Marketing in Only 2 Minutes a Day: Youtube, Google, Amazon, Cross Promotion, Blogs, Shapr Book 2.'

Correct Answer: If you named Plexo or Astore, you learned well.

7. What is Plexo?

Correct Answer: Plexo is an Amazon tool similar to Adsense. It can be used to create products similar to the keywords on your website including graphical ad links.

8. What is Astore?

Correct Answer: Astore is an Amazon website you can use on your own to include only those products you want to promote on a web page with their prices, pictures and a brief description. If you have a website, you can place Amazon ads in it and when someone makes a purchase you get your own commission. You could also place its ads on your blog.

9. How often does Amazon pay its Amazon Associate Program members?

A) Weekly

B) Bi-weekly

C) Monthly

D) Every first of the month

Correct Answer: Amazon pays on a monthly basis through check.

10. How much money can be made on a sale?

Correct Answer: You can make up to $20 to $100 per sale on Amazon depending on the product you are promoting.

11. True or false? If the person you, the Amazon Associate member, referred did not make his first purchase on the first day, you still get your commission if the customer purchases with 60 days because of Amazon's 60-day cookie feature.

Correct Answer: True.

12. True or false? Amazon Associate members must use the links Amazon provides in order   to get your sale credited appropriately.

Correct Answer: True.

13. Which of the following is NOT one of the ways suggested in Book 2 for promoting Amazon's products.

A) PPC

B) Article marketing

C) SEO

D) Blogging

E) Classified ads

F) Squidoo marketing

G) Social marketing

H) Pfishing

Correct Answer: H (Pfishing is not recommended as a way to promote Amazon's merchandise)

14. Book 2 suggested a cool way to market Amazon products with success by conducting thorough product research. After which, you write a review on that product and submit it to article directories. Why is this a recommended practice to make your money?

Correct Answer: Such a practice facilitates getting targeted traffic from search engines, and also from the article directory due to your article being indexed by people searching for them.

15. Why is writing up the products you promote with review articles worth the effort?

Correct Answer: These days people always search for a review on a particular product before they make their purchase, and if you are able to write a good review, you may get a couple of sales from just one article. Besides, making money online with Amazon is a very lucrative business once you are able to play your cards well

16. What is included in a comprehensive article on the Amazon products you promote?

Correct Answer: Include the pros and cons of that product in order to make the
reader know more about the product and also include the reason why
your reader should   consider purchasing the product.

Amazon Ads

You have choices regarding how to monetize your website. How you want to be profitable crushing social media using Amazon ought to be your considered decision.

Choice #1: Adsense

1. How do you make money with Adsense?

Correct Answer: AdSense program pays you a certain amount of money for each advertisement that is clicked on your website.

2. What sense is there in deciding to make money through the Adsense program?

Correct Answer: Consideration is determined by a number of factors and can range from anything as little as a few cents to as high as several dollars per click. While you can't determine exactly what you will make for each click, you can make a good guess based on the market that your website is based on. For example, a web site based around car insurance is sure to pay out much more per click than a website based on can openers.

3. What are two primary determiners for getting paid well by Adsense?

   A) the subject of your website

   B) the amount of traffic you can deliver

   C) low click-through rates

   D) location of Adsense ads on your web page

   Correct Answers: A (the subject of your website), and B (the amount of traffic you deliver), and   D (the location of the Adsense ad on your web page which get you the best click-through rates)

Choice #2: The Amazon Affiliate Program

1. How do you make money with Adsense?

Correct Answer: The program pays you a certain percentage for each sale that is made within a 24 hour period of someone clicking on a link from your site.

2. What is the Amazon Affiliate Program's Upside?

Correct Answer: You receive credit for any item bought through your affiliate links, not just   the particular item that was being advertised on your site.

3. Name another benefit to the Amazon Affiliate program.

Correct Answer: It is perfect for websites dealing with certain products, as you are able to link directly to products that you may be discussing on your website using pictures rather than just text links.

4. Which website monitizer should you use?

Correct Answer: If you said 'it depends,' or perhaps, 'both.' you are correct on both counts. If   your website deals with products, then you should be sure to use the Amazon Affiliate program   to promote the products you are dealing with.

If you are promoting an expensive product, you can be sure to make good money through   Amazon. However, there is certainly no harm in running AdSense ads on your site as well,   giving your visitors more options to choose from.

Many companies advertise their products through AdSense as well, so running these ads on your website can't do any harm. Leave it up to your visitors to make the choice. They will   choose what best suits what they are looking for.

Running both AdSense and the Amazon Affiliate program has proven to be the most profitable   way of monetizing for many intent upon crushing social media marketing.

How to Make Money Through Google

Anyone looking at generating income online and moreso through a work-at-home or freelance business, cannot afford to ignore this search engine giant. Despite the bad publicity from the many online scam sites posing as legitimate income generating businesses, there are many legitimate ways through which you can make money online.

How to Make Money through Google AdSense

1. What is the key factor to make money using Google AdSense?

   A) a beautiful website

   B) amazing information

   C) ads

   D) creating a heavy traffic flow

   Correct Answer: D (creating/attracting a heavy traffic flow)

2. How do you attract more visitors to your site?

   A) pretty images

   B) long-form sales copy

   C) use of search engine optimization (SEO)

   D) pictures of politicians

   Correct Answer: C (use of search engine optimization (SEO))

3. What can SEO do for you?

   Correct Answer: Optimizes your website around the most effective relevant keywords, or   keyword phrases that pertain to your site. (This simply means that you are making your site an   attractive option to the various search engines so that when someone types those keywords into   a search box, your site will be high up on the search-engine-generated results page list. This   increases   the   odds   of   someone clicking through to your site.)

4. What about the Google ads, themselves. How does one use them?

Correct Answer: These come in many different shapes and sizes. Some have photos; some   don't; some run the full width of your web page, while others are relatively small.

5. Does it take a lot of skill or nerdy knowledge to use Google Ads?

Correct Answer: One does not need to put a lot of thought into the ads themselves since they   will appear automatically after you know how to invoke them.

6. Why would you want Google to place a Google search bar on your website?

Correct Answer: If used, it brings the visitor to another page with even more ads.

7. Can opting in to the Google Referral program also make you money online?

Correct Answer: Yes.   The Google Referrals program pays you to bring in new AdSense   members.. If someone visits your site and clicks an AdSense link and signs up; when they earn $200 from Google AdSense, you earn $200 too.

8. Google AdSense is the easiest way for a beginner to make money with Google, but what can   it really do for you? Check any box you believe is correct:

☐ works through your own blog or website

☐ it use is free, so, there is no money to pay to opt-in

☐ you use the embedded code issued to you to link to Google's massive database of mini ads and it generates ads that are relevant to the content of your site and the page that each visitor to your website is on

☐ you receive a payment from Google each time a user clicks on any of of the ads

Correct Answer: If you checked all the boxes, you got them all and did not miss what was   presented in 'How to Crush Social Media Marketing in Only 2 Minutes a Day:

Youtube, Google, Amazon, Cross Promotion, Blogs, Shapr Book 2'

9. If Google AdSense does all that for you, what's left for you to do?

Correct Answer: Google only places the ads on your website and commits to pay you for each   click. Your goal must be to attract as many visitors to your website in order to increase the odds   of users clicking the ads.

10. How does one increase visibility of his website?

Correct Answer: There are a number of ways through which you can increase you website's   visibility including back links, social bookmarking, generating articles or content that resonates with your target market and many other search engine optimization techniques.

Google Alerts

If you have a need or desire to keep track of what is going on in the news, as well as on authoritative websites and blogs, Google Alerts will keep you in the know.

1. That's nice, but how does Google Alerts make you money?

Correct Answer: Quite simply because Google notifies you of any material information on your   watch list as soon as it emerges. You can then use this information to develop fresh content that   engages   visitors   to   your   website   and combine this with the AdSense program to generate   revenue.   Engaging, attracting, sending visitors is your best shot at making money online.

Google Search

The goal for the internet marketer is to keep eyeballs on his business website as long as possible. You want lingerers. This is one time when loitering is a a good thing. In effect, you want your site stickey. Google can help. The reason why this tool is effective for publishers is this. A visitor has been searching around for something which lands him on your site. Unfortunately, you haven't got the information he wants so he needs to continue searching. If you haven't got a search box, the visitor will almost certainly return to Google to carry on with the research. However, if your web page embeds its own search engine on your site returning results from Google, he will probably use that and also very probably click one of the links which comes up. That's how to make money with Google search.

Google AdSense Tips

Tip #1. Keywords:

You should care mightily for the importance of keywords. Pick two reasons from the list below:

A) pertinent keywords create payout of the ads

B) keywords are drivers for your site's ranking in the search results

C) keywords are over-rated

Correct Answers: A (pertinent keywords create payout of the ads) and B (keywords are drivers for your site's ranking in the search results)

2. True or False? Powerful pertinent keyword means higher rankings, higher rankings means   more visitors,   and more visitors means more clickability.

Correct Answer: True.

3. Proper keyword selection means high payout rates. True or False.

Correct Answer: True

4. There are two sites of many on the web that can help you find the right keywords for your   site. Can you name the two mentioned in 'How to Crush Social Media Marketing in Only 2   Minutes a Day: Youtube, Google, Amazon, Cross Promotion, Blogs, Shapr Book 2?

Correct Answer: Overture Search Inventory, Google Keyword Tool. Both keyword tools will   give you a list of related keywords, and their popularity.

5. The Google Keyword Tool offers a Keyword Density Analyzer. How is that important to   you?

Correct Answer: It also offers the possibility to check estimated payouts.

Tip #2.   Write many informative, high quality, niche pages

This tip comes as a two-parter. Part one is the more pages you have, the more AdSense   banners   you can put up, and the more visitors you can gain, and the more returning visitors you   can   gain (due to the changing content).

The second part is make sure your webpages have added value bringing new information to the   reader. Do not copy pages from other sites, as Google will recognize them as being a copy, and this will eventually lead to fewer visitors.

Page design and grammar matter: a pleasing format, not overthrown with colorfull flashing   banner and pop-ups and popunders. high quality, nice and readable, a good writing language, no   grammatical, syntactical or spelling errors. More languages means more visitors, so translation to French or German will generate more money.

Tip #3. Don't over use CSS

1. Using CSS is good. Using CSS is great. Using CSS really enhances the looks of your   websites, so why is overusing CSS a bad thing?

Correct Answer: Overdoing CSS can damage the relevance of your ads, and thus the high   payout rate of your ads.

2. What damage will overusing CSS do?

Correct Answer: Google AdSense reads your text and your HTML and recognizes tags to determine which keywords are most important on your website. Different tags get different   ratings by the Google AdSense Brain. Google AdSense needs the full palette of keywords to be   associated with your site, so do not throw it off so that it does not post the most relevant ads. If you use CSS to format these parts of text differently, the Google spider will not be able to   make   a   difference   between   the   different keywords.

Tip #4.   Don't put Google AdSense on an empty page

Remember, do not put AdSense on an empty page because it throws off Adsense, which cuts into your money making. Suffice it to say that this is bad for the combination of Google   Adsense and Google Spider tools which together determine relevant ads to display for your site.   No one wants a vastly diminished payout rate.

Tip #5.   Use Channels

Using your Google AdSense homepage, create channels, and use them to follow up on the   popularity (AdSense wise that is) of the banners on your different pages. Name your channels and your banners descriptively for tracking and analysis purposes. Knowing how they perform tells you which object earned you which money.   Use this information to tweak your pages, your   banners,   your   placement,   and   your   banner colors. The Click Through Rate (CTR) of your page   is mostly dictated by those four items.

Tip #6. Positioning Ads

Positioning your ads is very important and not as straightforward as one might think.

1. What are two considerations for ad placement?

      A) the Google Heat Map developed after study by Google showing the best positions for ad   placement

      B) ads above the fold line generally perform better

      C) ad placement is more science than art

      D) ad placement is more art than science

Correct Answer: A (Google Heat Map. They gathered all their stats, and figured out that when all other things remain equal, the placement of the ads gave this performance map.) and B (above the fold line   placement)

2. What did the Google Heat Map reveal about the use of colors in ads?

Correct Answer: The color varies from dark orange (strongest performance) to soft yellow (weakest per   formance).

3. What else did the Google Heat Map reveal about ad placement?

Correct Answer: Ads placed near images perform better, as do ads placed near control items (menus).

4. Which of the following violate Google's ad placement regulations?

A) placing images in the ads

B) placing ads in such a way that the visitor may be confused by closeness of a link and an ad

C) drawing the visitor for clicks using pointing fingers, arrows, etc.

D) placing an ad beneath the fold line

Correct Answers: A (placing images in the ads), and B (placing ads in such a way that the visitor may be confused by closeness of a link and an ad), and C (drawing the visitor for clicks using pointing fingers, arrows, etc.)

Tip #7. Colors of your ads

1. What opinions are there regarding colors of AdSense ads?

Correct Answer: Opinions differ. (1) Blending in your AdSense ads with the rest of the page gives the best results, while others say (2) making them in totally different colors, (3) making them with borders   will attract more clicks to your ads. (4) Blending the ads into your webpage means using the same style,   (5)   use   the   same background color for the ads as the background of your webpage, (6) make the text color the same as the text color of your web page, and (7) do not use a border (not using a border is done by setting the color of the border to the same color as the background of the ad and of your webpage).

2. What did the book suggest?

Correst Answer: Blending them in surely has the nicest effect on your pages, it doesn't really mess up the overal look of your website, and seems to be less annoying for your website visitors. Blending the ads into your webpage means using the same style, use the same background color for the ads as the background of your webpage, make the text color the same as the text color of your web page, and do not use a border.

3. What did the book say regarding using channels in respect to colors of ads?

Correct Answer: Again here the rule comes in to use channels, use good names for your channels, and experiment a lot.

Google Adwords

A Google AdWords campaign will bring the right type of visitors to your website.

Youtube, Google, Amazon, Cross Promotion, Blogs, Shapr Book 2'

By Ndeye Labadens with assist from P. Sterling Underwood https://www.amazon.com/P.-Sterling-Underwood/e/B005FCC9BG

About the Author

Ndeye Awa Labadens, DBA Candidate, is an author, entrepreneur and world traveler. She is a DBA candidate in Business Administration with a specialization in Entrepreneurship.

Follow her:

Twitter handle @labadensndeye

Instagram handle: Labadensndeye

Sign up for a free e-book/Online course copy here:

http://www.lannconsultings.com

Or email Ndeye Labadens ndeye@lannconsultings.com

And for a How to Crush Social Media in Only 2 minutes a Day newsletter sign up here: https://app.convertkit.com/landing_pages/146362?v=6

https://www.goodreads.com/user/show/57638047-nan-labadens

Goodreads groups

https://www.goodreads.com/group/show/209363-lann-group

https://www.goodreads.com/group/show/224746-review-seekers-group

LinkedIn: https://www.linkedin.com/in/ndeyelabadens/

OTHER BOOKS BY NDEYE

Australian Memories - Discover Aussie Land and the Mysterious Red Center

Relocation Without Dislocation: Make New Friends and Keep the Old

Secrets Book Launch Journey to the Ultimate Success Book 1

African Memories: Travels to the interior of Africa

European Memories: Travels & Adventures Trough 15 countries

Playbook Strategy: Stand Out Like a Business Giant

Asian Memories: The Delights and Discoveries of an incurable traveler American Memories: Cross Continent Adventures and Discoveries from North to South

Secrets Book Launch Journey to the Ultimate Success Book 3 - 6

How to Crush Social Media in Only 2 Minutes a Day: Twitter, Facebook, Kred, Goodreads, LinkedIn

Available here: http://www.lannconsultings.com

Resources

Facebook groups

Promo Giveaway

https://www.facebook.com/groups/571752053022689/

Infopreneur at

https://www.facebook.com/groups/197779700625069/

Ndeye's Book launch

https://www.facebook.com/groups/291763474501493/

Facebook Pages:

How To Crush Social Media In Only 2 minutes A Day

https://www.facebook.com/HOW-to-CRUSH-Social-MEDIA-in-only-2-minutes-a-day-143943039672975/

Ndeye Labadens fan page at

https://www.facebook.com/ndeyelabadens

Coachings at https://www.facebook.com/Koaches/

LANN Consutings at

https://www.facebook.com/LannConsultings2016/

LANN Consultings for interviews and coaching help toward your journey

http://www.lannconsultings.com

YouTube:

Testimonial:

https://www.youtube.com/watch?v=BtSl4l22w1l&t=1s

How to Crush Social Media

https://www.youtube.com/watch?v=mrPGzY7ueOg&feature=youtu.be

RIPL.COM

https://www.youtube.com/watch?v=pChl1T8_dV0

https://www.youtube.com/watch?v=dXfyTnyuE98&t=1s

LinkedIn:

https://www.linkedin.com/in/ndeyelabadens

Articles and interview samples:

https://www.linkedin.com/pulse/vlogging-2016-digital-marketing-challenge-labadens-dba-candidate

https://www.linkedin.com/pulse/save-your-children-stop-abuse-ndeye-labadens-dba-candidate?trk=mp-reader-card

https://www.linkedin.com/pulse/death-december-shonah-stevens-ndeye-labadens-dba-candidate?trk=mp-author-card

https://www.linkedin.com/pulse/suffering-from-your-neck-back-painthen-interview-sean-ndeye?trk=mp-author-card

Get your keywords using Buzzsumo: Keywords Buzzsumo and list of influencers and Social Quant link here https://www.youtube.com/watch?v=MicstRMv7C8&t=1s

Goodreads groups

https://www.goodreads.com/group/show/209363-lann-group

https://www.goodreads.com/group/show/224746-review-seekers-group

Join me at: https://www.goodreads.com/user/show/57638047-nan-labadens

After your Influencers research you have their email, or Twitter account and other social media to reach out them. Set a strategy to reach out to them and inform them about your book soon to be released. If it is already out then offer them a copy in order to get more reviews or opportunities.

Join the Facebook group at

https://www.facebook.com/groups/ndeyelabadensinterviews/

Bestsellers Giveaway promo group be seen by Influencers and other bestsellers.

Join the group and Share your book in promo links with authors and E-readers at

https://www.facebook.com/groups/571752053022689/

Coachings is my Facebook page where, Ndeye, share some coaching tips about growing your audience. Join here: https://www.facebook.com/Koaches

Evernote

I am not a great fan, but if it is your cup of tea, it happens to be a useful tool.

The Email Game

If you feel overwhelmed then this tool will help you manage your email in a timely manner.

Vocaroo for voice mail recording

MailVU

For video email saving you time of typing and editing before sending your message

Rapportive for Gmail: Social spying at its finest.

Rapportive shows someone's latest updates on Facebook, LinkedIn, and Twitter to the right-hand side of a message they've sent. It also helps you make sure you're connected with the people you want to be and shows a photo of the person of whom you are corresponding.

Wise Stamp

Wise Stamp is a fun tool for creating a fancy signature in your emails account. You can use it to promote your author page if you have many books. If you have single book, then add your book link or website.

Convertkit Manage your emailing list use Convertkit. I recommend it for the rules and the best time saving. It deserves the time to set your campaign.

http://mbsy.co/convertkit/27430407

Everlessons

If you ever considered having courses I would recommend you this Everlessons course https://jvz1.com/c/809511/269743

Payment method

Use JVZOO https://www.jvzoo.com/register/809511

Paypal

https://www.paypal.com/uk/invite?token=6CpBj1zinUc&program_code=Signup_Referral
_Notification

Copyrights

Copyright © 2017 Ndeye Labadens

www.lannconsultings.com

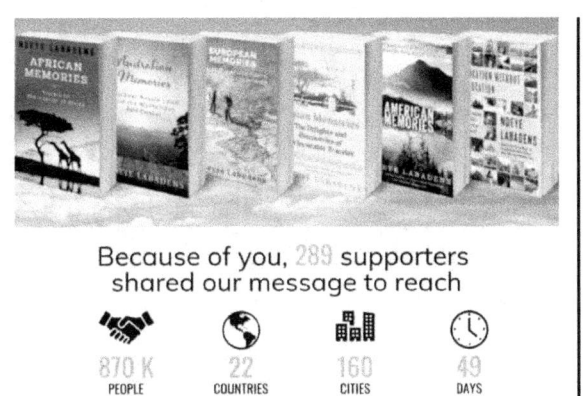

Because of you, 289 supporters
shared our message to reach

870 K	22	160	49
PEOPLE	COUNTRIES	CITIES	DAYS

Type to enter a caption.